For Family:

Dad, Mom, Stephen, Christine, Monica, and Kevin...

For Teachers:

Joanne, Kenny, Brice, Deborah, Linda, Bruce,

Mark, Tara, Brandon, and Dagen...

For Friends:

James, Chad, Doug, Anne, Christopher, Dave,

"sgJunior," Peggy, Reid, and Dylan...

For Therapists:

John, KTL, Martha, Carol, and Laura...

...And for our Four-legged Friends:

"You guys have saved my bacon..."

(and more than once.)

A Dog Ear'd Cosmos

Second Edition

David Mitchell Boie

A Dog Ear'd Cosmos

ISBN: 978-0-6151-6367-3

"Woof!..."

Thompson??!! ...recognize that bark anywhere...

Yo dough-boy...

But this seems even crazier than I'm accustomed – what with the history of schizophrenia -n- all...

And a pretty weak beginning for a book... but how else can a dog announce his presence?

I guess you have me there... having some doubts about "the book thing," though...

Just open up, listen up, have a piece o' patience pie, and witness what unfolds...

But if you remember, old' 'Teaser, sometimes Dave here was pretty severely negligent with you...

You talks funny... But hey, that's water under the bridge, 'old friend... fair warning though, as you listen with this pup, the guilt you entertain tempts you into considering I could harbor resentment against you...

Listen, D, be mindful when listening for the Angel of Puppy-Dogdom with me, for the Unconscious has the goods on you, and if you ain' vigilant – indulging in guilt-laden states – the ego takes hold the pen and mimics my voice, only with diabolically cruel intent.

So I should "center-up" before and while listening with you, with Faith that you have released all grievances, as puppies are accustomed...

Correctamundo.

So... now what, ol' pup?

Well, here you sit, just beginning to open up your awareness with the many friends (canine and otherwise) who have been waiting for you to welcome our offers of help.

Yeah, guess there has been some fear around sharing some ideas that could open a few minds, depending on how folks receive them...

David, this be your 'ol friend, "Thompson" coming on back with you (also referred to as Fuzzy-Beast,

Teaser, Thomper-Lou, T-willy-O'mpkins – and that's just an abbreviated listing of the family nicknames for the 'ol dog) – you can breathe easy. How these ideas end up being received jus' ain' your call.

That's actually a relief...

Yeah, you have perceived yourself as isolated and have been fearful of sharing fairly powerful ideas on the edge of awareness – regardless of what you have learned to the contrary, you still identify as a source of ideas, or at least as a kind of craftsman of ideas.

You have been trying to "go at 'er" in this world, and you have shunned the simple help and faith you were prepared to accept as a child. You even dither over silly things like word choice and punctuation, thinking that if you make even a single critical error, the world might come crashing down.

What vanity you indulge in!

Strangely, I feel good hearing you say that...

Rejoice that you lack power to injure the authentic Self with your mistakes. In holding mistakes lightly, you have all the power and help you need to perform your small part in bringing wholeness into the world. Rejoice, also, that your part is fit perfectly for you, as is everyone's.

...in starting this little "project," I had tried to begin by writing a "heady" intellectual essay as though old Dave-wave was the source, and wound up bogging down in doubt...

Any helpful writing simply comes as the Light of Wholeness enters your mind... and if you find help in asking Spirit to communicate with you as though a beloved childhood pet were intervening for you, then we shall adapt to your needs...

Thanks... the old Mind appears to be "settling down" now.

Nothing that helps you put fear to rest could be beyond Spirit's willingness and power. And yes, you may consider that "Dog be your co-pilot." In your case, you learned to care and play with a dog by your family's side.

Thanks, again... I feel more at ease.

David, you have washed up "going on your own."

You see fear and animosity lurking around you, though you refuse sharing openly - couldn't these be "interconnected phenomena?" Come on!!

When you put the matter like that...

In writing, you are doing your best to communicate honestly – that shall be felt... and heartfelt. The opening has come, and you have waited long enough to be embraced.

As a child, I remember dreaming of being respected in the world as a physicist or inventor.

With just enough "mathiness" and pop-culture to listen and hear in terms that both scientists and laymen alike may resonate with – together we can offer a vigorous "deep-tissue" massage for your intellect.

Specifically, we can massage certain "blind-spots" in the Western World view you inherited that have imprisoned your awareness. All who walk the earth appear blind to some degree – and consciousness tends to paper-over our blind-spots such that we become unaware of even having them.

> *...also remembering having had a science book for kids that included some exercises demonstrating the blind-spot in the human visual field, corresponding with the location where the optic nerve connects with the retina in back of the eye – this was a weird experience for me, and one that had a lasting*

influence.

Perhaps drawing awareness into some of science's blind-spots, this dialog can facilitate a realization: you do have choice in whether to reinvest these insensitive and numbed areas with the healing power of Self.

Science has had a huge intellectual hard-on for a Grand Unified Theory, a "Theory of Everything..." and you, having a fixation with science and substantial, yet still limited math skills, have continued fondling this extension of the public ego with an unwarranted awe – mixed with jealousy.

Yeah, I have entertained jealousy toward the big scientific players – at the graduate level, seems I jus' had difficulty with "the math."

When humans deflate this intellectual erection through loss of interest, our spiritual power can flow toward a healthy and more balanced recognition and nurturing of simpler, more openly powerful gifts of healing – much to the detriment of those favorite emotions our collective ego seems to thrive upon: jealousy, envy, pride, despair, and outright hatred.

Whoa, you jus' went off, dude...

Ain' even done... as the collective Mind becomes less convoluted, our spiritual gifts shall appear simpler, more child-like, and in that measure they become more like the gifts of Christ.

I hope so. Hey Thompson... kinda' seems hard talking about some of these blind-spots;

they seem largely contextual, and jus' ain' part of the conventional dialog. Ain' what science says that seems off kilter; more like what she ain' saying.

David, as a dog, though I may have been color-blind, old Thompson could sense things that appeared invisible to you, and offer you a gentle heads up.

So, where do you see our scientific blind-spots highlighted in what ain' being said?

Well, perhaps the largest error and blind-spot in physics appears in the tendency to extrpolate data and theories well beyond what appears a sensible scope.

In pursuing a Grand Unified Theory, science attempts to extrapolate human subjective power right out of our official picture of the world.

...Such that the vitality of people (latent, expressed, and repressed) appears minimized in potential and import.

Yes, in large measure, pursuing a Grand Unified Theory involves the attempt to find Laws governing the very smallest of scales... Laws which might extend in their predictive power and apparent validity, "explaining" the immensely large. In the middle ground, the power of our human choices involving forgiveness get the squeeze, appearing subsumed and squelched – "caught between quarks and curved space."

What lies behind this scientific attempt to "predict" everything?

In part, ego-based fears seem involved. For starters, don't get fooled by the idea that the probabilities of quantum mechanics (or a detailed scientific understanding of how they "work") grant anyone room for living happily.

Quantum Mechanics offers no real respite from the old view of the Universe as being a giant "clockworks" with cruel gears that grind human beings into angered and despairing fragments.

We do have room to stretch, although mastery of the physical world does appear an illusion. Rather, observing with the Silence of God (as best we can) appears as the only answer that can soften these

"cruel human gears."

So where can we go from here?

How 'bout a roundabout introduction to the principle of "Conservation of Energy" – as violated by contemplative practices?

That sounds a little far-out there... Could we discuss another example of how science tends toward vanity, and remains ignorant (or at least unwilling) to question her own presumptions? Maybe we could eventually work our way around toward "Conservation of Energy" that way.

OK... as a first case, consider how physicists ascertain that the universe "must" now be expanding. Some review might be appropriate –

you remember witnessing how the pitch of a train horn goes up in frequency as she approaches the listener, and then drops as the train passes...

> ***Yeah, in Newtonian physics, that's the Doppler Effect – the frequency of the horn's "sound" appears different to an observer who moves with respect to the train. As the train approaches, the pitch appears "artificially" elevated, and then when she passes and suddenly starts going away from the observer, the "effect" operates the other way.***
>
> ***And so the person standing on the side of the tracks "hears" a sudden drop in pitch. Meanwhile, on board the train, as the theory goes, the horn would sound normal the entire time – constant pitch.***

And when galaxies or stars move away from an observer, a similar phenomenon appears witnessed known as a "red shift."

Yeah. According to current models of space-time – perhaps somewhat inaccurate although they are the best we currently have – "Red shifts" can occur as a function of a distant light-source having a conventional velocity that increases the distance between the observer and the star or galaxy over time, or, even weirder, "red shift" can occur as a function of the whole fabric of space-time stretching and expanding.

The later phenomena often appears likened to watching ink dots on the surface of an inflating balloon. As the balloon fills up with air, the rubber "fabric" stretches and the dots appear as being farther apart – even though they don' move within the surface of the balloon...

And judging by the fact that lights more distant in the sky appear to be receding faster then closer stars and galaxies, scientists conclude, based on some assumptions, that the universe is currently expanding...

Yes... though matters ain' as simple as this. Scientists also use data from far-flung supernova explosions, background radiation, and the like.

And physicists may very well be correct in

most of their assumptions – in fact, I suspect they are. Jus' seems like our physicists' overwhelming bias involves developing a model of cosmic evolution that minimizes, even eliminates, the role consciousness plays in manufacturing d' experience called "Cosmos."

And you have found yourself asking, what are those assumptions...

One thought that came into awareness was that the red-shift "data points" physicists rely upon for their projections reflect the state of stars eons ago. I mean were talking billions of years in most cases... at least for this kind of data.

And the "further" a data point appears in the sky, the older she be. So scientists appear to be taking various kinds of data from all over the space-time "map" and projecting onto a template reflective of their biases... and numerous templates appear possible, including some that might account for the universe "unwinding" as a function of human spiritual evolution, or at least having this potential.

And this kind of extrapolation has seemed presumptuous, from your perspective?

Well... there have been changes in the meantime. I mean, billions of years ago, supposedly our solar system had jus' begun evolving.

So if a hypothetical human-like race in some far-distant galaxy were to assess the earth's condition "now," based on "measured astrological data" millions (or perhaps even billions of years old), the assessments made of earth would appear distorted by the appearance of nothing but a big old ball 'o lava in the viewfinders of their advanced imaging telescopes.

Touché...

And, as physicists have said, based on Einstein's equations for general relativity: matter, energy, space, and time all appear related and interwoven. So if space-time can appear as being "pushed-out," "pulled-in," and "twisted-up" like macramé, then according to what limited sense I have, something "other" than space must be willfully "doing this."

Yes, though even now, physicists' overwhelming desire appears to revolve around establishing a model that could account for these transformations without human consciousness having any role to play.

As Einstein's metric equations have indicated, space-time can have varieties of tension, pressure, and (via convoluted mixtures of these) a potential for "twisting" or "knotting-up" into entanglements of numerous sorts. Each of these factors appears represented in Einstein's brilliant General Relativity Theory (perhaps overly brilliant, and thus, limiting in some strange ways).

Your job in "writing" this ain' to be some new "Einstein" type person – though this idea does appeal mightily to your sense of vanity. (Wink – wink.) We likely ain' gonna be developing any brand new theories or math-based "predictions" that can be verified by some earth-moving scientific observations. Einstein did his job, and did his job well enough.

Your job, David, centers around addressing your Friends trapped in the Western World view from an informed perspective. Together we shall look into what most Western scientific theories have in common: an ignor-ance of the human subjective power of both interplay and "interbeing," to borrow terminology from Vietnamese Zen Master Thich Nhat Hanh.

The theories of science appear powerful, though Science stands upon a bedrock of human goodwill that she often betrays. Perhaps rather than using their brilliance to squeeze human subjectivity "out 'de picture," scientists could begin helping people develop skills in handling their own gifts of subjective power with appropriate responsibility.

> ***OK… so then, circling backwards – who, or what, does the "weaving" of this tangled Cosmic web – where Subjectivity appears "split" into multiple perspectives who long for Union, and yet appear so confused that our very attempts, and even our perceived desires, run counter to our real goal?***

OK, Davey-boy, looks like you 'wanna "jump on in..."

...the Subjectivity you share with your fellows has dreamed this Universe... and you folks be the "Dream-weavers." You may look on this Cosmos, or "the fabric of space-time," as a veil you have collectively woven, and hide behind.

Much as in the allegory of Plato's Cave, a sudden exposure to the fullness of "da Love" would overwhelm you. Hence the Universe shall have a gradual "undoing," as gradual as you, your Brothers and Sisters choose to have – and all under guidance and direction of the feeling humanity has dubbed "happiness."

When a veil of time like this Cosmos appears woven amidst Eternity, the World of Forms appears as an imaginary divide separating our Father (the "giving" aspect of Self) ...from the Son (the "receiving" aspect of Self) – And, though you might "think" you wish otherwise, the Son appears as having responsibility for undoing this veil.

In Eternity, all Joy is shared perfectly and instantaneously, therefore "to give and receive are One." And so, no divide between Father and Son appears in Heaven.

For the time unbecoming, however, you do perceive a veil of separation... an entire world of your own collective making... and you have a temptation toward despair at the prospect of taking responsibility for the undoing. Give thanks for your Brothers and Sisters, for we offer help.

This feels kinda' intense... could we circle back to a discussion of science's view of this, jus' to ground this dialog a smidge?

Officially, science as now exists would have difficulty with much of this...

How's that?

When Einstein expressed concern over the published results of the "EPR experiments," experiments that seemed to provide confirmation of theories of quantum entanglement, he complained that the results validated "ghostly action at a distance."

In some measure, science appears a protection against conscious recognition that the Magical World view we each experienced in childhood indeed does have veracity, and, in fact, has deeper roots than the Scientific World view of the West.

From where does science inherit this bias?

In early childhood, the world does appear magical to infants and toddlers, and we don' necessarily mean this in a nice, "story-book" fashion. Most people who develop the intellects necessary to excel in scientific endeavor do so as an unconscious strategy of defense against the darker aspects of "a magical world."

The perceived emotional dangers of growing up as a child in some emotionally numbed families often acts as a spur to a child's intellectual development. First, the child seeks to understand how parents could be so blind to the importance of the emotional-intuitive world, and then, in process of developing the intellectual skills to communicate across the divide, the child loses sight of that world herself.

Indeed, the sharper ones' intellect (or "mental body," in Hindu terminology) becomes, the more effective that person can seem at "winning" arguments, debating, etc. Or, in short, invalidating people – a variety of "dark magic."

What could this possibly mean within context of a maturing Worldview?

Well, if a "white magician" uses words and "non-verbal" communication to make his fellows feel good about themselves, and a "dark magician" uses words and "non-verbal" energetics to encourage a person to doubt their own value and embrace the darker aspects of fear, then most "people," intellectuals included, seem drawn to both... often in completely inappropriate circumstances.

How could "white magic" be seen as inappropriate?

Well, even "white magic" poses some difficulties in many contexts, as the ego strokes involved seem to encourage a sense of co-dependence and reliance on "externals." Occasionally, such strokes appear needed for encouragement, although these "strokes" can also appear highly addictive and can tempt people to continue looking outside themselves for "the Joy that abides Within."

I guess I don' quite see how Science necessarily embraces "dark magic."

Regardless, aspects of Modern Science may soon appear in hindsight as a form of "ambivalent alchemy"– where "the ends" are thought to justify "the means." This kind of orientation implies insensitivity, and this particular brand of insensitivity packs a pretty big punch.

Science's paradoxical pride in (and placing of value on) her very claim of being "value-free" – this amounts to a form of advanced Orwellian "double-speak," albeit unconsciously motivated and deeply misunderstood.

In dismissing philosophical questions of "value," scientific spokesmen gloss over what seems a profoundly important issue – and in light of the fact that through science, humanity has the power to incinerate whole cities, perhaps we might consider that the very concept of a "value-free" knowledge amounts to a diabolical contradiction in terms.

How we experience the world appears constrained in no part greater than by the qualities of beingness we value... and invest in.

> *Science seems to push detachment as being helpful. Ain' "detachment" often posited as being a useful spiritual skill?*

With regards scientific detachment as compared with Buddhist detachment, this seems like a case of two different birds going by the same name.

Could you elaborate on that, Thomps?

OK. Though "offered up" as a rationalization for continuing on with a cold-edged interrogation of nature, "scientific detachment" amounts to a choice of, and a vote for, continued emotional and intuitive numbness – a deliberate callousness laced with barely acknowledged aggressive tendencies.

Science, as a collective enterprise, compulsively seeks to extend the ability to predictively man-handle the world remotely from within the Ivory Towers of the Intellect...

...when denied access to particle accelerators, biotech laboratories, and other scientific tools, the "playpen" of the intellectual shrinks to a desperately small size – and then, psychoanalytic terminology like "object-loss" can seem far too tame of labels to accurately represent the intensity of consuming despair.

Does science have any authentic spiritual roots?

Of Course. Modern Science began in the crucible of Newtonian Observation... the observer who could sit back, contemplate nature, and align his ruminations with the rhythms of the Cosmos such that he might receive understanding.

Nowadays, the mental vortex spins even faster, and many scientists don' seem to find their lives centered in contemplative stillness, but rather feel like they have to struggle jus' to keep afloat in a very competitive intellectual climate.

Egos' and arguments are tough and toughly fought within the Scientific Academy. Advanced science appears to have, in large part, devolved into a form of male heroics, a "grail quest" of denial where the violence we commit and are subject to in our daily comings and goings (usually verbal and emotional) "doesn't matter a whit," so long as we get our paper on cosmology, QED, "String-Theory," or what-have-you written and published to rave "peer-reviews."

So, the scientific mentality has been abdicating responsibility for this lack of values?

Yeah, pretty much. Again, these barely conscious denials of responsibility amount, upon closer inspection, to "affirmations of despair."

Science's alternatively despairing, aggressive, and hubristic attitudes have been encoded within scientific language as the Fundamental Laws of Frustration: Conservation of Energy along with the Laws of Thermodynamics.

Our current "Fundamental Laws" of Physics appear related with a fixation on explanations that minimize the possibility of Awareness, or Subjectivity, having any capability or willingness to find rest on occasion, allowing this twisted and knotted-up Cosmos and linked Mind-set the much needed opportunity to "unwind" – so as to offer more space for experience of simple "beingness."

From a quasi-Platonic perspective, before the Divine Reunion, our Universe must first "lighten up," unwinding like a rubber-band into a happier configuration: a saner world where the prospect of recognizing Eternal Innocence appears less daunting. David, remember that miracles just ain' currently in the allowable purview of modern science – which appears obsessed with willful control and manipulation of external circumstances.

Uh... Thompson, though I don' mean to sound overly flippant, how could "miracles" even legitimately be spoken of in scientific terms? Isn't that anathema?

Yes, anathema to some. Although for those who have ears to hear: Miracles basically "amount" to subtle, yet powerful disappearances of matter, space-time, and associated guilt-laden thought.

The Laws of Thermodynamics, conversely, have

a decidedly "anti-miracle" bias, emerging, as they did, out of "thought experiments" highlighting the limits encountered within the design of conventional Heat-engines during the formative stages of the Industrial Revolution.

Highly accurate within context of the scientific assumptions taken to be "true" at the time, these "Laws of Frustration" express the idea that, left entirely to herself, a mechanistic Universe would, theoretically speaking, wind up experiencing a "cosmic-crucifixion" of sorts – referred to in science as "the Heat Death of the Cosmos."

And what about Conservation of Energy?

If this Universe (composed of varied forms of energy) appears "hellish" in some time periods, and merely mildly-purgatorial within others – then the Law of Conservation of Energy, in most basic of terms, amounts to a highly intellectual way of saying: "this Nightmare World appears permanently inescapable."

Weird dude, sometimes the scientific drama appears deeply poetic... if also tragic.

You hinted earlier that science appears as an extension of the collective "public" ego... are scientific attitudes mirrored in each of our private psyches?

Yes, in some measure, we all be li'l scientists in our own worlds. Sizing up the competition, addrssing problems and, more authentically, opening our minds on occasion to larger perpectives, possibilities, and

more potentially liberating ideas.

The ego appears – in "milder" forms – as a generally defensive orientation towards the World... and, in more virulent forms, as downright hostile (and even murderous) attitudes.

Negative "ego-attitudes" appear embodied in the human organism as the habitual "ocular tension," dubbed "ditension" by researching psychiatrist Trigant Burrow. Most of the time, people look at one another – their work – and even on the objects of their surround as if they were confronting a dubious problem, puzzle, or adversary of sorts.

Science pushes this dubious attitude toward the World in the extreme, having "sprung from" the general populations' coming further and further

under sway of this secular and worldly orientation.

How does one begin loosening the grip of this trap?

Spiritual practices like Meditation, Contemplation, and various "thought replacement" practices of some Hindus, Jews, Buddhists, Muslims, Christians and, more recently, students of A Course in Miracles – each of these manage to disengage Subjectivity from identification with the ego's thought-stream... while concurrently "teasing out" a relaxation of the human organisms' ocular tensioning patterns – meaning the habitual tendency to "pull" a sharp, steely-eyed focus on the objects of perception.

With a bit of disciplined spiritual practice, a softly

warm peripheral-focus begins to develop, allowing for "the World" to come into (or out of) awareness as a holistic gestalt, rather then as a scrutinized "figure" against an ignored "ground" – to use the jargon of Gestalt psychology.

> *I have been playing dangerously with alcohol consumption recently, and wonder how drug and alcohol intoxication play into all this. I have used them to "stuff the gap," though their abuse seems to "short circuit" any dawning spiritual awareness via a false promise of "Love in a Bottle."*

Many drugs that folks use to "get high" seem to mimic the effects of disciplined spiritual practice – filtering out ego-based thoughts via a "forced" relaxation of our ocular tension and the linked association with point-focus. With a few drinks, the world seems softer and our tendency to view her through problematic eyes wanes somewhat.

Booze and drugs have well documented pit-falls of their own, and a temporarily "happy drunk" must come down from his artificial-high and "face the music..."

When people drink, they often do so to "shut off their minds" and quell unsociable, "crazy" thinking. The forced ocular relaxation provides a taste of what unitive experience feels like – a state devoid of "racing, crazy thoughts."

On "thoughts," could you elaborate jus' a bit more?

Considering that medical science appears able to measure "brain-wave" patterns with EEG electrodes resting lightly on the surface of the human skull, the idea that "thoughts" are confined and exist exclusively inside the brain appears somewhat preposterous.

In actuality, the brain, nerves, acupuncture system, and perceptual apparatus function like a pattern-able "antenna array" that typically has been trained to "pick up" thoughts consistent with our particular personality structure.

When one considers that the "thought-forms" these antennas resonate with pre-exist, are fundamentally impersonal and communal in nature, and are immanently blended into and throughout the physical world – the idea that thinking is "in here" and the World appears "out there" tends to break down somewhat.

Saner living, then, becomes a matter of "tuning in" to healthier thought-patterns associated with healthier contexts and groups of people. Alcoholics, for example, seek refuge in the Fraternity of AA, and often have to learn how to dissociate themselves from old "drinking buddies" without generating internal strife.

Our "thought-antennas" function not only individually, but in group dynamics – sort of like the powerful "grouped antenna arrays" used to strengthen and direct receptivity in television and radio broadcasting.

Where do "thoughts" come from, and how

are they either "reinforced" or, alternatively, have their influence "undone"?

Thoughts appear expressed in terms of language, and language appears as a social phenomenon – hence thoughts be communal in nature. We reinforce "thought-forms" via our habitual patterns of speech and via our emotional reactivity to spoken words and non-verbal communications with our fellows.

Thus, negative thought-forms may find themselves "disappeared" or "grounded-out" through non-reactive and compassionate listening... and this could account for the (as yet somewhat limited) successes of therapy and psychoanalyis in changing our linked patterns of unhealthy perception, thought, and verbal interchange.

...and for the sake of comfort, could we at least pretend to tie all this back into science and physics?

OK. Thoughts appear as a form of subtle energy with an as yet undetectable structure. They move in "orbits" extending both through physical and "non-physical" space.

By non-physical space, we refer to the multi-dimensional volumes enclosed by curved space-time. Due to mental blocks and blind spots in our collective awareness, few physicists have bothered asking what may soon seem obvious questions...

...Like, for example: "If space-time does appear curved, then what could she possibly 'curve-around' or enclose?" Could the "Dark" Energy postulated by physicists correspond with "what lies beneath" the perceived surface of our Universe (or the "stuff-n-nonsense" that curved space has, to a relatively large degree, kept enclosed?)

In the corridors of Academia, quantum particle interactions have been thought, since Richard Fenyman's work, to somehow be "a-causal" – as the Fenyman diagrams describing them appear to hold validity regardless of a particle diagrams' "orientation" with respect to the chosen space-time coordinate system.

If these types of interactions are viewed as being undulations in the surface of an ocean of repressed awareness – with Dark Energy correlating with "what lies beneath" the perceived surface of this *Cosmos* – then scientists may find many of the mathematical quandaries and conundrums associated with things like "Dark" Matter and Energy, the Zero-point of Einstein, "Free-Energy" phenomena, and many other of such mysteries "resolving" to an extent, and perhaps sooner than had been anticipated.

David, when you identify exclusively with "the body" and with the "thoughts" associated with your particular personality-structure, you are basically functioning as a "flat-lander." This amounts to saying that, when centered in the compulsive judgments and deadly seriousness of an "ego," you have a greatly restricted awareness, and may find help pondering ideas like this: "Could a 'flat-lander' re-discover that his World actually be 'curved?'"

While viewing this world through defensive eyes, people invariably find themselves looking to understand and exploit what seem like causal relationships in the World of experience. However, our typical "experiences" reside only within the "surface currents" that flow upon the very top of this ocean of repressed awareness.

In moments of what Carl Jung called synchronicity, the apparent causal associations we had made between perceived patterns find themselves "undone." These associations often "wash away" in many varieties of non-ordinary experience. And so, some of the flawed "causal" connections we had postulated are thus revealed as having been largely "epiphenomenal."

David, the realm of "Cause" resides in the Abstract Mind ...in wordless "Aha-moments" that abide outside the realm of Form. The power of these moments "manifests" as relaxations in the unfolding patterns of the Universe, via changes in our habitual patterns of "thought" – both spoken and unspoken – as well as corresponding changes in our sense of "the lightness of being" and overall emotional tone.

When you find yourself painfully busy 'reacting,' 'anticipating,' and 'strategizing' your "next move" in the game – based on what you believe you have accurately learned to see and understand as a typical human being saddled with "five senses" and a judgmental human intellect – well, then you're very much like a dog chasing (and often times cruelly biting) her own tail.

Since this applies to people everywhere, the "Dark" Energy (or Unconscious Mind) that sleeps fitfully beneath the perceived surface of our physical world may appropriately be seen as

"collective." We are all in this "boat" together, and there are some very Loving Beings with far greater spiritual power and responsibility than you could currently handle watching over you. Remember this.

OK...

...seems like, with human consciousness typically constrained to the "turbulent surface" of our Cosmos, most of the people operating in this world wind up feeling cramped for "psychic space" and find themselves embroiled in battles over personal territory, both physical and mental.

Yes: And what with experience unfolding like swirling iridescent patterns in the surface of a "Cosmic soap-bubble," when things ain' quite

“spinning” your direction, you folks tend to react defensively, feeling unduly threatened by the idea that your individual “bubbles” might burst, leaving you lost in an unmanageable chaos.

David Mitchell (as “Mom” addresses you when in trouble), you often seem “pissy” in circumstances like these, and you justify your bad feelings based on what has just “happened to you.” However, this puts the “cart before the horse,” so to speak. In actuality, your decision to “play” with negative attitudes comes, in causal terms, *before* any perceived “external irritations.”

And so your ingrained habits surrounding the indulgence of “sour-grapes” attitudes (as well as other “attitudinal flavors” currently featured in our Ice-cream and Sundae Parlor of Misery) – these indulgences present an immeasurably larger threat to your Peace of Mind than any resultant form-level triggers to which you may be reacting.

Remember Dave, the perceived physical, emotional, and mental movements within the patterns that comprise the experienced Cosmos (and which erroneously appear to humanity as being "the whole ball 'o wax") – these patterns are, in fact, only "surface storms" on a subphenomenal Sea of repressed Awareness. And you can learn to "ride 'em out" within the safety of Peace of Mind.

> *Hey Thompson... if we may "Go Greek" for a moment, sounds like these "iridescent patterns" of experience correspond well to what Plato described as "shadow-plays" – with "the cast of characters" in our drama including quantum particle interactions, "macro-scale" objects (like planes, trains and*

automobiles), and even human feelings and thoughts.

Yes. And while contemplating what a mind-blower this may seem, remember that you don' necessarily need any of this information to function compassionately in the world. You are now "living" in very much the same kind of Dreamworld you were as a child, a naive college student, and even as a "schizophrenic" – and yet you managed to get along pretty well without this info.

And so, when you feel overwhelmed, odds are you're trying pretty darn hard to get "one-up" on the Cosmos and "fix things" Solo-style – all by yourself – without the help of you and your Brothers' and Sisters' connection with the Spirit of Peace.

If you hold these complicated questions about physics lightly and loosely, the "answers" we find together may prove helpful ...especially if this discussion serves, in part, to draw some attention away from the "big game" of Science and naturally "re-directs" this attention toward the "middle game" of exploring the subjective power within Everyman.

Anyhow, getting somewhat "back on track" here – we were discussing how the subtle energy of "thought-forms" may be disappeared via attentive and engaged listening.

With regards Cosmology, these kinds of "disappearances" of varied forms of matter and energy, however subtle (and corresponding local variances of the "shape" of space-time) could effectively account for the differentiation of objects and experience within our world.

When considering the "Big-Bang," physicists have been puzzled as to why the universe has developed differentiated structures, emerging as we have in a spontaneous, non-dimensioned (and therefore symmetrical) "Singularity." What could account for the breaking of symmetry?

We inhabit what seems a dynamic Multiverse, rather than some single static solution of the "metric equations" of Einstein. And considering the theoretical symmetry of our emergence from any Big-Bang "Singularity, " these differences appear easier to process and integrate if we allow for the possibility of some matter and energy having "disappeared."

And a "stepping back" toward an acceptance

of responsibility for one's own experience – as well as dis-identification with ego-directed "thought-forms" – practices like these account for such a lightening in the "weightiness" of the Cosmos?

Yes, seasoned with an appropriate sense of humor and playfulness. Having fun appears quite integral to "lightening up." Within any World view that minimizes the role of Subjectivity, such decreases in the weightiness of the Cosmos would appear hard to account for – although the ego appears bound to encourage a continued thrust toward a de-spirited view of mankind.

The Cosmic "soap-bubble" idea would be analogous in the String Theories of physics to a "World-sheet" ...shaped like a multi-faceted "soap-bubble" and buffeted about by "internal winds" of various sorts. Within human consciousness, particles and objects appear to interact causally as the surface of this "soap-bubble" undulates with the equilibrium-seeking movements of "what lies beneath."

Within the surface of this "World-sheet," some happenings (or events) may, for reasons of practicality, illegitimately be called "causes" of other events – although this would appear to be like watching shadow play on a wall and trying to say "shadow A caused shadow B to do X, Y, and Z."

The fundamental difference appears as one of "shadow vs. substance" – and even our unconscious motivations ain' "Real" in the sense of being Eternal... as the Universe opens up to the Light of Innocence, both consciousness and unconsciousness resolve into the Divine Awareness of Pure Being.

And, although your awareness may appear temporarily constrained to "the surface of things," the Freudian vision of liberating the Unconscious (or slowly letting the air out of the balloon) still appears as a useful metaphor.

How does Freud mesh with other psychological perspectives, including Game Theory, Transactional Analysis and other stuff like that?

For the most part, the kinds of "games people play" involve attempts to expand personal territory within the surface of "the Cosmic bubble." Or, perhaps more concernedly, humans aim to negotiate time-binding exchanges where perceived burdens are shifted around from one set of shoulders to another throughout time and space.

These exchanges appear arbitrated as agreements, commitments, contracts, and expectations (often unspoken) - all such that individuals are promised a tolerable amount of experiential "energetic flow" breathing in and out of their personal worlds.

Mystics, conversely, seem more practiced in contacting (and aware of) the river of Joy that flows "perpendicular" to the walls of what is conventionally recognized as "real."

Perhaps this requires being able to have the boundaries of the personal-self appear softer and more flexible, while maintaining awareness of our Brothers' and Sisters' own boundaries... like learning how to “surf” the wave of the Unconscious - delighting (particularly) in an appropriate lessening in the suffering of our fellows...

Delving into the personality structure a little, if we may, how does the fascination with individualized personal identity develop?

Through “identification” (me, my, mine) with certain of the “iridescent patterns” we experience swirling around the surface of this shared Cosmic “soap-bubble” (cars, bodies, trinkets, experiences, and thoughts) – and through rejection and judgment of other patterns – our attention seems seduced and “hooked” into constructing a personal identity, and instead of a delightful costume-mask party, we end up in a madhouse prison of mistaken identities.

Perhaps our healing could involve a soulful internal recognition: we are most helpful as stewards, both of what we "own," and of the history of karmic merit and integrity we have earned and nurtured as personalities struggling to find our way amidst an intensely confusing world.

Bingo, baby!!

Our "identity boxes" appear defined, in large measure, by likes and dislikes (fairly obviously)... although also by the correlations of ourintentional efforts directed toward achieving goals and our subsequent witnessing of the fulfillment (or frustration) of these goals, as they appear embodied within the unfolding patterns of this Big ol World we have made.

A difficulty in authenticity seems implied by the

idea that many of our goals, and the linguistic distortions we use to symbolize them, may also simply be embedded in the "iridescent patterns," along with the outcomes of our applied efforts.

We thus may regard ourselves (from a limited human perspective) somewhat as programmed "flat-lander's" – listening to and identifying with fear-based egoic promptings that appear embedded (or "scripted") within the fabric of space-time, herself.

> *Like watching a very dramatic 3-D virtual reality movie, and getting all wrapped up in the plot... including the "internal soundtrack?"*

Yes, precisely. Dave, in responding reactively to the movie – or what Buddhists call "unmindfully"

– you appear fixated and preoccupied with "the surface of things."

...And quite literally so... for the sights, sounds, emotions, and "internal monologue" of the perceived World appear as the surface of "a bubble of separation" – an imagined separation from your Brothers and Sisters and, hence... from the very Source of Life – to Whom the concept of fragmentation would be incomprehensible.

Illusory or no, sometimes, this kid feels overwhelmed by fear that God will take away my "separate identity" before I am ready. And sometimes I feel like I don' do all that great a job at releasing grievances and my attachments to the things of this world.

You ain' expected to give up worldly ambitions

until you are thoroughly ready. The perfectionistic desire to "skip steps" along the path amounts to a cunning "ego trap." So be gentle with yourself, homeboy!

Furthermore, the Comforter (or Higher-Mind) ain' out to "burst anyone's bubble" – and the concept of "force" appears completely alien to Spirit. You are perfectly welcome to reject the ministrations of Spirit, though the rejection of His Comfort appears, well... increasingly uncomfortable.

The Spirit of Wholeness bears no blame for the suffering of the World, and in the final "judgment" neither do you... you and your Brothers and Sisters are Perfectly Innocent of blame, yet appear nevertheless responsible for your part in redemption from illusion.

This may seem off topic, but could we address my fascination with the drama of the Zero-Point Energy discussion happening on the internet?

Remember, D, the intricate details of what appears to happen within the surface of the Cosmic "soap-bubble" ain' really what's pivotal, cause you have little, if any, control over that stuff. In many instances, appearances may seem deceptive... you are reminded to look within for Joy.

Miracles of forgiveness are "the order of the day," so to speak. And the Universe may, as science reconsiders the inviolability of "Conservation of Energy," appear more and more to humanity as being a kind of "spiritual cramp" within the Mind of God.

Zero-point Energy has seemed fascinating for you from the standpoint of "Free-Energy," but more significant appears the ability of beings to disappear matter and energy via attentive listening – joyfully "holding our tongues" for a Holy Instant, and letting Spirit "pull" healing words from us.

This, in Toltec wisdom, be called "The Power of Silence."

Humanity "takes turns" using the Atmic "perspective-point" (or Zero-point) - from which we collectively manage to project a vastly addictive experience called: Cosmos. And although beyond the scope of current physics, spiritual traditions the world over acknowledge that humanity inhabits a fractured and "fallen state."

Until scientists embrace and integrate mystical experience, what science has to "say" shall appear biased in the extreme by rejection of such teachings.

So Thompson, I should like to "confess" something...

Ten-four, 'ol buddy... *go on ahead...*

Well... in practice of scribing this dialog, I have felt like I'm just making the communication up. I 'still sometimes shift your words around to suit where I think things should head, and I guess that has me pondering...

Yes...

Seems potentially unfair to present this writing as a dialog...

And so...

And so, have I just been talking with myself here... doing a kind of "creative writing" exercise...

And you believe that's what's going on...

At some level I really do...

Well, look at things this way... dialog feels authentic only when you are prepared to listen and be changed by the process of engagement. At the level of authentic encounter, the brittle flavors of our individualized personalities begin to soften in the warmth of the shared Light.

So what if this "creative writing" exercise, as you say, sounds somewhat like you on both ends, I guess I would ask you: Which side has brought you more comfort?

Well, your side, far and away... that's where the unexpected departures have occurred from the plot line I had already established in my mind about the way this should go.

In giving you freedom to speak via me having willingness to listen and surrender some of my desire to control the dialog... that's where the unexpected flashes of light have come shining through.

You gosh-darn-skippy about that. Jus' remember the value of even a little willingness to be childlike in faith and open to the Holy Spirit's comfort. Yes, these messages appear all jumbled up... you often put words in this dog's mouth that ain' there... in fact, that's the case with the Holy Spirit in many instances.

You might consider just remaining in meditative stillness and attempting to "bliss-out" all the time, saying nothing, hearing no thought-stream, etc.

And that can be a great refresher, that contemplative experience of stilling, slowing down, stepping back, broadening focus, and looking inward – or as they say in the movie "Bull Durham" ...breathing through your eyelids.

I dug that movie...

There just ain' need to hold an inflexible insistence on internal silence being compulsive...

in Zen Buddhism, that's called "Zen Sickness." Warm, forgiving, grateful thoughts have great power in many contexts.

So if you ain' necessarily Thompson...

The Voice that brings comfort is far more Thompson than you realize... and Thompson is far more you than you care let into Awareness...

Well... guess I ain' gonna deny the power of this "creative writing" exercise...

Just remember, what you believe for now simply (and literally) appears before you as what you believe for now.

You may regard this as a writing exercise if you wish... either way, if your willingness to participate in the “exercise” has changed your patterns of perception in the direction of healing, then large sections of space, time, and associated guilt-laden thought have disappeared (or at least begun unraveling).

And, if you had remained unwilling to humble yourself and look “childish,” this “lightening-up” would have had to wait, or would have happened far more slowly and painstakingly.

> *So I’m OK, even having placed some unsubstantiated physics ideas in your mouth?*

Well, consider... how could this really matter? Miracles ain’ subject to being taken back or damaged, as all they “amount” to are disappearances of space, time, and guilt.

So when you flounder around listening to the ego's BS, you ain', like, damaging some precious storehouse of irreplaceable antiques called miracles that can never be recovered. You're simply spinning your wheels.

Hmm...

And remember, David, everyone works miracles every second of everyday, continually, or rather, on a continuingly intermittent basis... you just haven't learned to work miracles habitually and consistently.

Huh? So like, in any given second, the typical ego-bound person performs somewhere between, maybe like, a few and a couple hundred miracles a second...?

You may sound like you're being sarcastic in saying that, but actually that's a fairly conservative guess. These miracles appear worked inadvertently, almost despite the conditioned thought patterns you have.

That's how Holy we are. In the depths of being self-programmed to experience misery, we occasionally forget the deadly seriousness with which we take our struggles. In that brief instant, some of the "massiness" or "weightiness" of space-time disappears. We "lighten up." We remember our salvation.

> ***Does the ego try to use those moments of relief toward furthering our sense of separateness?***

Remember, the ego ain' really able to claim the

power of a miracle, for a miracle don' amount to anything. Miracles ain' doings or makings, but rather undoings.

The ego does attempt to convince the Self that separation brings real comfort, though the ego's game becomes more and more transparent as our collective "guilt-mass" lightens up...

...and the sense of what your mentors have called "the feeling," "the flow" or "lightness of Spirit," the sense of relief we experience whenever the painful spiritual cramp that appears as this Universe opens up to the nourishment of lightness... this "feeling" appears as the Comfort of the Holy Spirit.

You mentioned the phrase "self-programmed for misery" earlier...

And you have some first hand experience with programming computers. Dave, we have awareness of where you want this to head now.... fair warning though, this may be beyond your ability to "get across" effectively – especially considering you speak English, one of the most "high entropy" languages in the world.

What do you mean by "high entropy" languages?

Before looking into that topic, you might consider how distorted your views of the relationship between gratitude and happiness have been...

...as a corrective, you may consider what people call "happiness" as a function, kind of like a

mathematical function, of gratitude. Gratitude is what we most fundamentally are, and happiness flows from a stepping back toward a recognition of what we are.

This may be a tough pill to swallow, but people don' develop ungrateful attitudes because they are unfortunate, or feel dissatisfied or "unhappy," although there appear powerful reinforcing "negative-thought patterns" that occur in these states... very difficult to surrender attachment to these modes of thinking. No, rather people manifest unfortunate circumstances as a function of their historic karmic identification with thoughts that encourage an ungrateful orientation.

Hmmm...

This may seem a tautology, but consider: have

you ever remembered "being" ungrateful or identifying with ungrateful "thought-patterns" in the midst of feeling happy? Do you really consider that happiness appears as some magic car that drives you out of the ungrateful attitudes and thought-identifications you had previously been indulging?

Whoa... sometimes, I've felt happiness flowing after long periods of silence, work, or "alone time," and then when I have interacted with people, have experienced a crash of sorts.

Maintaining a grateful orientation requires vigilance – and if you consider that you "feel" or "are" thankful *because* you're happy, and engage in interpersonal dialog with this idea, you appear likely to "short circuit" your dawning remembrance of your grateful nature in short order.

David, this whole sick ball 'o wax appears in large part as a case of the wolf thinking the tail is wagging us – and even further, at a deeper level, of the Innocent Spirit of Subjectivity considering the Self a wolf.

Oh, the man-y layers of subterfuge.

That ain' the half of Id... and that's OK for now.

Again, you spoke of English being a "high entropy" language...

A friend of yours, Kenneth Pratt, was looking into how "percussive" English, German, and similar sounding languages appear, with all their “ch’s” “ck’s” “k’s” “t’s” and “cht’s”. If we grant any significance to the Power of the Spoken Word, humanity may gradually wind up abandoning some “phonemes” (or “sounds”) over time as language morphs.

How so?

Well, English and German, in particular, seem difficult to “speak softly in” and tend to “Disturb the Peace.” You have noticed how people seem more vocally animated in Spanish and French speaking countries, and speak louder when arguing than we do in the United States, Great Britain, and Germany... Oui?

Oui...

And you’ve pondered that French and Spanish

appear as “softer languages,” with less percussive sounds, or “phonemes,” and that people seem to have to argue louder in “soft languages” in order to make the same amount of “noise,” or emotional impact, yes?

Yeah...

Well, Spanish and French appear, at least from the viewpoint of information theory, as having what might be referred to as less-weighty “spectral power densities,” or “phoneme bandwidth profiles” inherent in their spoken “sound.”

These “tongues” appear inherently less “noisy,” and people who speak these languages really have to “turn up the volume” to have the same emotional impact. You have read some on the

Albigensian Heresy...

Yeah... apparently, a spiritual "opening" occurred in the latter half of the middle ages, centered in the South of France around the town of Albi. Miracles and healings (some of them overly flamboyant, and "way out there") appeared along with a mixed bag of other stuff.

And perhaps the Buddhist ideal of compassion went out the window... the movement became somewhat gnostically "manic" in flavor, and the Albigensians' seemed to forget that we're all in this boat together.

Northern peoples seemed denied enough of a chance to open up and soften their egos gradually... perhaps dismissed as "emissaries of darkness," their violent reaction appeared embodied in the Albigensian Crusade.

And you feel curious about the appearance of softening accents in written French, and the dropping of percussive ending consonants on words in French speech that happened around this period.

Uh-huh... though I don' want to get seduced unnecessarily by "neat-looking" connections like that – guess I 'figure that the distortions of language can take care of themselves. Our job revolves around letting go of grievance... the rest can jus' unfold.

Excellent...

Oh… seems like, in some kind of manic excitement, the Albigensians could have forgotten the virtue of patience, and rallied prematurely around the old revolutionary (and Gnostic) cry: "Freedom Now!!!"

And yet I sense this misses, somehow – especially since many of the spiritual "leaders" of the Albigensian Heresy (also referred to as "the Perfect") apparently had far greater spiritual power than I would even want to have in this life...

Gettin' down on ol' Dave, now, huh...? Again... excellent... good work. Jus' 'cause you have chosen to avoid further foolish explorations of the seductive power that a newly discovered feminine receptivity has in the midst of such a repressed culture, this ain' necessarily an indication of

cowardice; outright insanity ain' necessarily "fun."

"The Perfect" appeared aware of the dangers they faced... they jus' may have "pulled on the envelope" a bit too hard, in a context where circumstances were even tougher... you may rest easy in the assurance that no miracle can ever be "lost."

Kinda like in Star Wars, when Obi-One lowers his light saber in the face of Darth Vader's rage, he left an influence with Darth... and Darth eventually was able to see that the shadows that had seduced his own awareness had no power to soil or sully the Light of His Innocence...

Shadows appear, rather, as Light turned back oneself... as Innocence resisting oneself; in other words, as a severe spiritual cramp, or intensely painful spasm, bound to exhaustion in the very effort to make “guilt” real.

The best policy for you around this time period revolves around moderate risks... a very brief moment of listening with the silence of Spirit before answering queries, rather than serving up glib, habitual answers when in conversation... jus' this shall liberate enough of the Unconscious for now.

Further, continuing work at the hardware store until asked to serve in new ways should be helpful for you... living on the dole has alienated you from people’s daily round of suffering, and your previous lassitude has aroused the workin’ man’s rage...

...some workin' folks have no idea the subjective hells you have been through while convalescing during this period of disintegration and incubation... and so, your having been on public assistance can incite repressed anger.

Be mindful of this...

OK. Hey, Thomps, just ponderin:

Uh...oh..

Yeah, well, after reading this, people may likewise consider: where does this leave "me?" How do "I" access the Light of Self?

That's a case of the ego talking... and you listening to "Id" – "both" in your case, an' in the case of anyone "else" who pays credence to the

ego and thus identifies Subjectivity with the "thinker" who be crafting ego-directed thoughts.

Paradoxically, we are encouraging you to experiment "playing" jus' a smidge mo' Self-centeredly... to experiment with a little Holy "Spanish pause" while looking inward before answering questions or returning greetings...

...To even experiment with "answering" a greeting with warm body language: warming eyes, a welcoming hint of a smile, a lovingly placed validation of beingness...

You have been used to giving habitual (sometimes convoluted, as the case may be) "ego strokes," unconsciously exchanged in measures designed to secure a sense of being one-up, in an effort to "come out on top."

Ain' this, like, encouraging people to play "mind-games" with one another?

This world appears as one colossal mind-game, and usually through complete identification with the ego's thoughts, the belief in irredeemable separation from God appears in charge.

We are jus' encouraging some room for exercises that shake your identification with the ego's thought-stream... but in process of that, awareness opens to the kind of violently manipulative games the ego has seduced us into complicity with...

Growing up, you considered that your own "ego," and perhaps to a lesser degree, those of your loved ones... were "innocent" while the "bad guys" were somehow separate from you... this ain' the kind of Innocence we be referring to.

This kind of "special innocence" appears as a great stumbling block to forgiveness and realizing responsibility for karma. Furthermore, in this world, "mind games" appear all that humanity seems capable of. This dialog appears as a "mind-game" also, although one dedicated in part to undoing Self-sabotaging strategies, and aligning "the work" with the Holy Spirit.

The ego has attempted seducing you on numerous occasions into a mute, despairing catatonia... and that jus' ain' 'gonna serve Innocence well.

But ain' Self-centeredness "mean?"

Face up, boy... the "ego" has the same "hellish"

experience “in Mind” for the “nice ego’s” you care for as well as those you dislike... the same atrocities committed, the same torturous suffering.

The “social pleasantries” of this world often appear as the antithesis of real kindness. When you and your fellows are being habitually “polite,” you are often identifying with a weakened, needy emotional “feeling-state” and are seeking an ego-stroke, or, conversely, social niceties often appear as mindless “brush-offs,” or even worse...

Sometimes, the ego encourages you into games where you might say “Bye, have a nice day...” and have a conscious intent of wishing someone a miserable day... you have been on both the “short” and “long-ends” of that kind of game, and can feel that kind of energy... as can everyone.

When tempted to despair, jus' remember... "the dreadful has already happened..." and do your forgiveness work... remember, people are hurting... aching to be genuine... so do your best, and no more... as don Miguel Ruiz has clarified.

Again, we ain' talking about being mean or harboring cruel intent while experimenting. We ain' talking about duplicitous intent, sarcasm, gossip, argumentation, etc. – all very Self-sabotaging games that you, David, continue to have an attraction to playing.

We are jus' looking into the possibility of playing with being "Self-center'd." Look into the phrase, David: "Self-center'd," or rather, "Center'd in Self..."

...when every perspective adopted by the Innocence of Christ appears similarly "Center'd..." – centered in internal Silence, Peace, and bathed in the Joy that flows in the Presence of Peace... well, then concepts like "when" and "where" ain' even 'gonna appear "sensible..."

You had begun, awhile ago, looking into the verbal content of interpersonal transactions as being like programming instructions for the Mind. Numerous fields of inquiry have considered ideas like this, including Scientology, Hypnotherapy and Clinical Hypnosis, as well as Neuro-Lingustic Programming.

Yeah, though I guess I had 'kinda had an "issue" concerning some of these fields. I still have some confusion as to whether anyone on this planet really knows what they are doing in these "meta-cognitive" fields... sometimes they seem more like potential "mine"-fields to me.

And just 'whatcha mean by that, home-skillet?

Well, speaking for myself as best I can, even though I have had some profound experiences of "opening up," I still lust after the female form, feel occasional jealousy towards my fellows, and generally would feel completely overwhelmed at the prospect of having someone looking to "me" to figure out their suffering and work to "reprogram" their modes of experience.

And yet this appears just what many therapists and NLP practitioners have either the courage or the foolishness to attempt doing. As a friend who started up a therapy practice which he later abandoned echoed: "fools rush in where angels fear to tread."

According to experts on Kriya Yoga, prior to the development of Sanskrit, Hebrew, and other "Mother Tongues," language wasn't used in the sense we are bound today. Rather, the seed-syllables of Sanskrit were used primarily as a "language of incantation."

As I would understand the matter, even though we have been saddled with virus-contaminated "high-level programming languages," like English, Spanish, German, French, etc... whenever we speak aloud we are still using and calling forth the immense Power of the Word: the pre-Sanskrit proto-language of mantras, spells, curses, and blessings.

And, even more daunting, we actually think we know what words mean, and that their meanings remain stable independent of context.

I daresay, David, you have really worked yourself into a 'tissy here. Language appears far more stable than you realize. And regarding your foolish experimentation with drugs, you appear saddled with a very hyper-sensitive mind... others among your fellows are, thankfully, far more stable in their perceptions.

All that you have said does reflect humanity's current situation, though you are forgetting one important fact: the split off fragments of the One Mind that appear to you as "other people" are your Brothers and Sisters... they are your Friends (although many, if not most, appear severely estranged from you).

Your unwillingness to see your fellows at least as friends in potentia (especially with your hyper-sensitive perceptual system) transforms many people into monsters before your very eyes. But consider this David, if you doubt them, how can you expect them to have faith in you?

In this holographic "house of mirrors," the external world does indeed mirror both your "personal" and "collective" internal frames of Mind, and though you may have moments of distrust, the amount of "good-faith" you have on deposit with your Brothers and Sisters acts as a ballast which you may rely on in times of faithlessness when you are tempted to "look into that glass darkly."

I don't mean to harp on you too strongly, David, for the Unconscious does appear pretty dark around this time period. However, indulging fear ain' gonna help you... to help you keep your courage up, consider that you live in a war-zone.

You can't hold on to your body forever, so you may as well use 'er up boldly in service of Love. Furthermore, your confusion about "the Power of the Word" may be an asset, as this can encourage you to deeply explore the "Power of Silence."

Words come automatically when needed and, often times, recklessly and compulsively when unneeded... loving silence requires discipline.

Ok... Thanks...

And David, remember that you're never really "alone" in the despairing way you think "alone" means. Even the objects you handle throughout your day have a kind of consciousness with which the Self can resonate.

David, you can unlearn what keeps you from being gentle with the objects, ideas, and bodies you touch with your senses and, on occasion, you may experience that this World be Christ's body... your body... everyone's body.

The Spirit of Innocence appears chained in every object and body you encounter (as the Hasidic Jews affirm) and the Joy of this Spirit can be

“tapped” and freed – with all due care and tenderness.

Zen Masters teach washing each dish as if one were washing the baby Buddha Himself... with a tenderness that inspires shared Joy. The Self, imprisoned by the failure to laugh with the idea of separation, does indeed appear as the baby Buddha.

And so – allow the Earth to express Her desires, frustrations, and Joy through everyone you encounter. The Joy shall intensify as this aching World and the “matter” of which composed are slowly and deliciously allowed to sublimate into the Pure Joy of Spirit – like ice off the sidewalk on a cold, sunny winter’s day.

And have patience, David, for the Holy Spirit ...if allowed... can transform even your biggest

perceived "mistakes" toward the service of God.

Thanks, Dough... Word up.